DIY SCIENCE FAIR FUN!

GEOLOGY PROJECT YOUR WAY

Megan Borgert-Spaniol

Super Sandcastle

An Imprint of Abdo Publishing
abdobooks.com

abdobooks.com

Published by Abdo Publishing, a division of ABDO, PO Box 398166, Minneapolis, Minnesota 55439.
Copyright © 2024 by Abdo Consulting Group, Inc. International copyrights reserved in all countries.
No part of this book may be reproduced in any form without written permission from the publisher.
Super SandCastle™ is a trademark and logo of Abdo Publishing.

Printed in the United States of America, North Mankato, Minnesota

102023
012024

THIS BOOK CONTAINS
RECYCLED MATERIALS

Design: Aruna Rangarajan, Mighty Media, Inc.
Production: Mighty Media, Inc.
Editor: Liz Salzmann
Cover Photographs: Adobe Stock; Mighty Media, Inc.
Interior Photographs: Adobe Stock, pp. 1 (hammer & rock), 5, 14 (clay), 21 (clay), 28 (clay), 29 (clay);
iStockphoto, pp. 4 (both), 6, 9 (river), 10, 14 (book), 15, 24, 26, 27, 31; Mighty Media, Inc., pp. 14 (baking
sheet, ball, marker, pebbles, bottle, ruler, sand, tape, towel, skewer), 16 (experiment), 17 (experiment), 18, 21
(pebbles, sand), 28 (pebbles, sand, experiment), 29 (pebbles, sand); Shutterstock, pp. 7, 8, 9 (boy), 11 (all), 13,
14 (measuring cup), 19, 22, 25, 29 (boy), 30
Design Elements: Shutterstock

Library of Congress Control Number: 2023939298

Publisher's Cataloging-in-Publication Data
Names: Borgert-Spaniol, Megan, author.
Title: Geology project your way / by Megan Borgert-Spaniol
Description: Minneapolis, Minnesota : Abdo Publishing, 2024 | Series: DIY science fair fun! | Includes online
resources and index.
Identifiers: ISBN 9781098292058 (lib. bdg.) | ISBN 9781098278953 (ebook)
Subjects: LCSH: Do-it-yourself work--Juvenile literature. | Geology--Juvenile literature. | Earth sciences--Juvenile
literature. | Rocks--Juvenile literature. | Science projects--Juvenile literature. | Science fair projects--Juvenile literature.
Classification: DDC 507.8--dc23

Super SandCastle™ books are created by a team of professional educators, reading specialists, and content developers
around five essential components—phonemic awareness, phonics, vocabulary, text comprehension, and fluency—to assist
young readers as they develop reading skills and strategies and increase their general knowledge. All books are written,
reviewed, and leveled for guided reading, early reading intervention, and Accelerated Reader™ programs for use in shared,
guided, and independent reading and writing activities to support a balanced approach to literacy instruction.

CONTENTS

EXPLORE GEOLOGY

Do you love to collect rocks along the shore? Do you wonder how rivers and mountains form? You might enjoy geology! Geology is the study of Earth's structure and features. Scientists who study geology are called geologists.

The most common type of rock in Earth's crust is igneous rock. It forms when magma from a volcano cools.

Geologists use rock hammers to collect samples of rocks.

Geologists **research** the soil, rocks, and **minerals** that make up Earth. They study the forces that shape our planet. They try to find out how land changes over time.

Geologists use microscopes to see what rocks are made of.

BECOME A SCIENTIST!

Scientists use a process called the scientific method. Check out the steps on the next page. You will use this method to **design** your own geology project!

THE SCIENTIFIC METHOD

1 ASK A QUESTION
What would you like to find out?

2 GATHER INFORMATION
What information do you need to understand your topic?

3 FORM A HYPOTHESIS
What do you think is the answer to your question?

4 EXPERIMENT
How can you test your hypothesis to find out if it is correct?

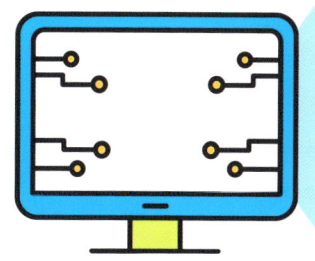

5 RECORD THE RESULTS
What did you observe in your experiment?

6 WRITE A CONCLUSION
Did your results support your hypothesis?

7

STEP 1 ASK A QUESTION

What topic do you want to learn about?

Maybe you are interested in caves. Or maybe you want to know more about erosion. Start asking questions! Write your questions in a notebook so you don't forget them.

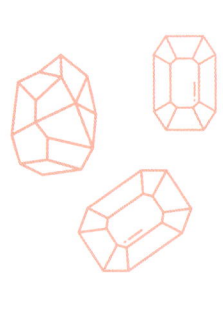

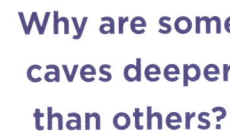
Why are some caves deeper than others?

How do caves form?

8

What is erosion?

How do riverbeds erode?

GATHER INFORMATION

Maybe you have a lot of questions. That's great!

Scientists often have many questions they'd like to answer. But for now, choose one to **focus** on. Save the others for **future** projects.

It's time to **research** your **topic**. You can gather **information** from many different sources.

▶ **Read online articles about the topic.**

▶ **Read books about the topic.**

▶ **Talk to scientists or other experts.**

What did you learn about your **topic**? Write it down in your notebook. Then you'll have all the **information** you need in one place.

Erosion is a force that moves sediment from one place to another.

Sediment is bits of rock, such as clay, sand, and pebbles.

Rivers carry sediment downstream.

11

FORM A HYPOTHESIS

After you research your topic, it's time to form a hypothesis.

Your hypothesis is what you believe is the answer to your question. First, revisit your question. Do you want to change it based on what you learned? Then think of a few different hypotheses. Record them all in your notebook.

QUESTION: Does the size of sediment affect how far it is carried downstream?

HYPOTHESIS 1

Larger sediment is carried farther downstream than smaller sediment.

HYPOTHESIS 2

Smaller sediment is carried farther downstream than larger sediment.

HYPOTHESIS 3

Sediment size does not affect how far it is carried downstream.

Now, choose which hypothesis makes the most sense based on your **research**.

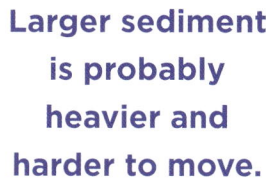

Larger sediment is probably heavier and harder to move.

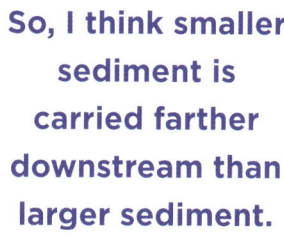

So, I think smaller sediment is carried farther downstream than larger sediment.

13

PREPARE YOUR LAB

Get ready to test your hypothesis.

Find an area with a sturdy table or counter to work on. Then gather the supplies you'll need for your science experiment.

SUPPLIES

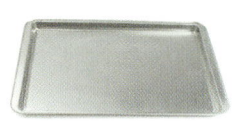
baking sheet with a rim

book that is about 1 inch (2.5 cm) thick

bouncy ball

clay powder

marker

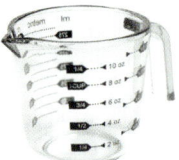
measuring cup

pebbles

plastic bottle

ruler

sand

tape

towel (optional)

wooden skewer

LAB RULES

All labs have rules that scientists have to follow. Here are some rules for your lab. They will help you stay safe and have fun while doing your experiment!

➡ **Ask an adult** for permission to use the materials and do the experiment.

➡ **Ask for help** with sharp or hot tools.

➡ **Wear goggles** and gloves to protect your eyes and hands.

➡ **Clean up** when you are done and put everything away.

STEP 4

EXPERIMENT!

You've gathered the supplies. You've prepared the lab. It's time to experiment!

1 Pour about 4 cups (946 mL) of one of the sediment types onto the baking sheet. Push all the sediment to one half of the sheet. Mark the sheet where the sediment ends. Then press the sediment so it slopes down toward the mark. Use the book to prop up the sediment end of the baking sheet.

2 Pour 1 cup (237 mL) of water into the bottle. Place the bouncy ball over the opening as a stopper.

LAB TIP

Place a towel against the empty end of the baking sheet. Or do the experiment outside or in a bathtub.

3 Hold the ball in place as you set the top of the bottle on the edge of the baking sheet. Make sure the opening is over the sediment

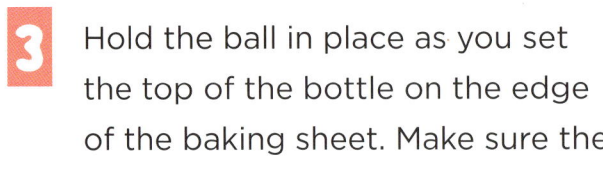

4 Hold the bottle as level as possible. Remove the ball from the bottle opening. Let the water flow from the bottle onto the baking sheet. This mimics the movement of water down a riverbed.

5 When the bottle is empty, notice whether the sediment moved past the mark. If it did, measure the distance between the mark and the farthest point of the sediment. Record the measurement.

6 Clean the baking sheet. Then repeat steps 1 through 5 with the other types of sediment.

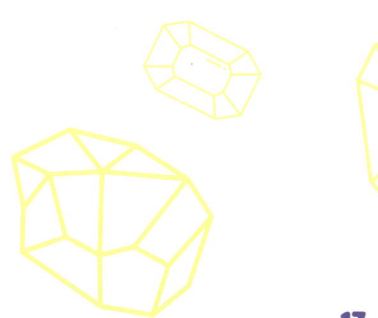

Look at the results of your experiment so far. You might be ready to draw a conclusion. But first, consider any other **variables** that might affect your results.

The slope of a river affects how fast water flows. Fast-moving water hits sediment with more force than slow-moving water.

You didn't want water speed to be a variable in the experiment. So, you made sure it was the same in all three tests.

I controlled the water speed by keeping the bottle level while the water flowed out.

And I kept the slope of the baking sheet the same in each test.

You also used the same amount of water and sediment in each test. So, the only **variable** was the size of sediment you used.

19

STEP 5

RECORD THE RESULTS

During experiments, scientists record data and other observations. You wrote down how far each type of sediment was carried. Now, it's time to record your data to share with others.

Scientists often use tables and graphs. This helps make the results easy for others to understand.

A table organizes **information** in rows and columns.

SEDIMENT	SIZE RANKING (big to small)	DISTANCE CARRIED
Pebbles	1	0.0 inches (0.0 cm)
Sand	2	2.0 inches (5.0 cm)
Clay	3	8.0 inches (20.0 cm)

A line graph shows the relationship between two **variables**.

SEDIMENT SIZE VS. DISTANCE CARRIED

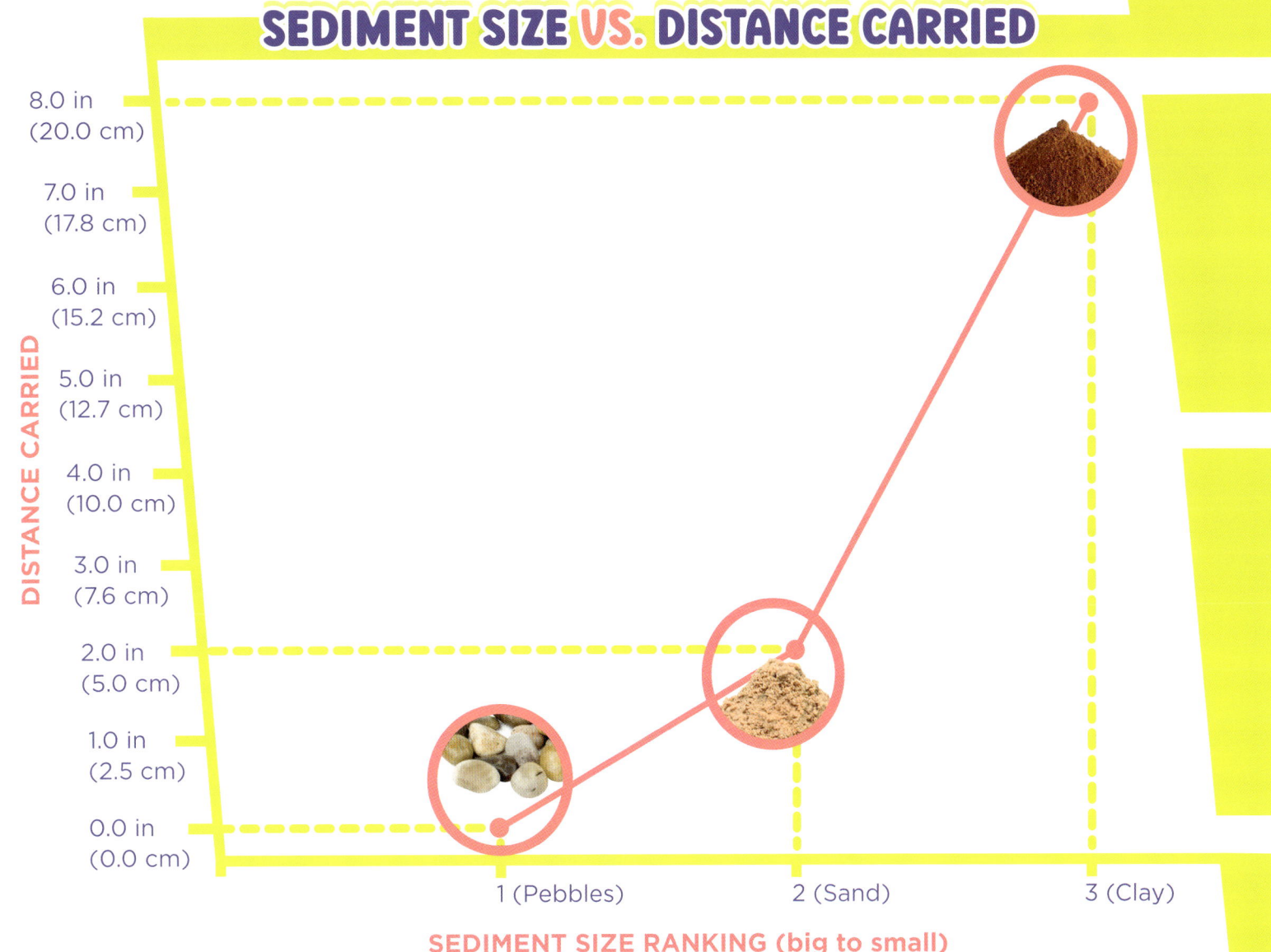

DISTANCE CARRIED

8.0 in (20.0 cm)
7.0 in (17.8 cm)
6.0 in (15.2 cm)
5.0 in (12.7 cm)
4.0 in (10.0 cm)
3.0 in (7.6 cm)
2.0 in (5.0 cm)
1.0 in (2.5 cm)
0.0 in (0.0 cm)

1 (Pebbles)   2 (Sand)   3 (Clay)

SEDIMENT SIZE RANKING (big to small)

WRITE A CONCLUSION

You recorded your results. Now it's time to write your conclusion. This is a **summary** of your experiment. Your conclusion provides the answer to your original question. It also states whether your results support your hypothesis.

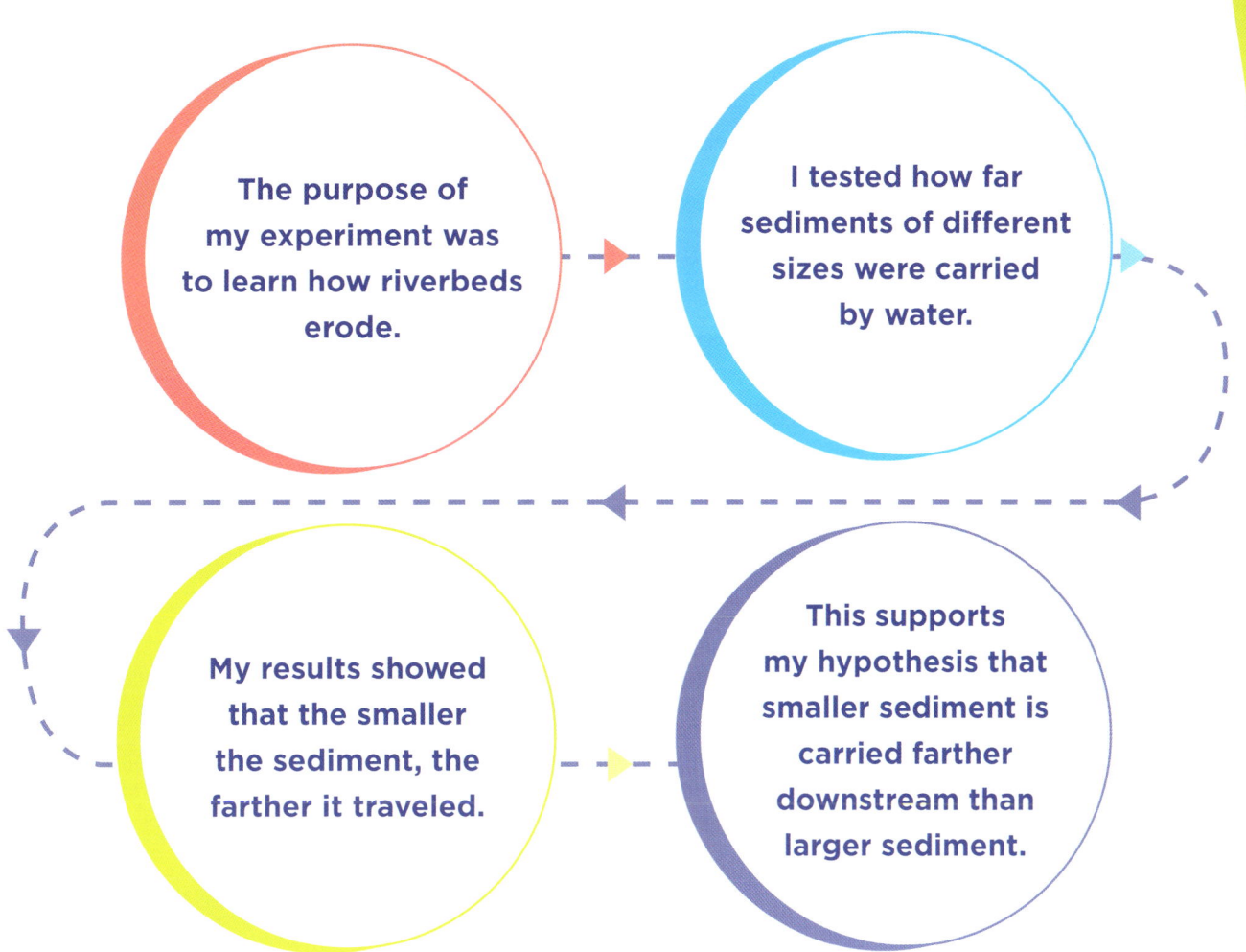

The purpose of my experiment was to learn how riverbeds erode.

I tested how far sediments of different sizes were carried by water.

My results showed that the smaller the sediment, the farther it traveled.

This supports my hypothesis that smaller sediment is carried farther downstream than larger sediment.

Many scientists find that their hypotheses were wrong. There is nothing wrong with being wrong! It means you did your experiment without **bias** and were surprised by the results. That is another mark of a great scientist!

FURTHER RESEARCH

A conclusion offers an answer to your original question. But it can bring up new questions too! For scientists, the end of one experiment often leads to more **research** and new hypotheses to be tested.

What new questions do you have? What could you research further? Do you have a new hypothesis to test?

Do plants affect erosion along riverbanks?

Does the slope of a riverbed affect how it erodes?

Does the shape of a riverbed affect how fast water flows over it?

PRESENT YOUR PROJECT

Young scientists share their **research** at science fairs. Students share what they learned with classmates, teachers, parents, and sometimes judges. It is a chance to show all the work they put into their experiments.

Demonstrate or show a video of your experiment.

Create comics or other drawings to show your project in a fun way.

Include props, models, or dioramas.

One way to present a project is with a display board. It should show how you followed the scientific method. Turn the page to see a display board of the project in this book!

QUESTION

Does the size of sediment affect how far it is carried downstream?

Pebbles

Sand

Clay

RESEARCH

Erosion is a force that moves sediment from one place to another. Sediment is bits of rock, such as clay, sand, and pebbles. Rivers carry sediment downstream.

RIVERBED EROSION

HYPOTHESIS

I think smaller sediment is carried farther downstream than larger sediment.

EXPERIMENT

The purpose of my experiment was to learn how riverbeds erode. In my experiment, I tested how far sediments of different sizes were carried by water.

Pebbles

Sand

Clay

RESULTS

My results showed that the smaller the sediment, the farther it traveled.

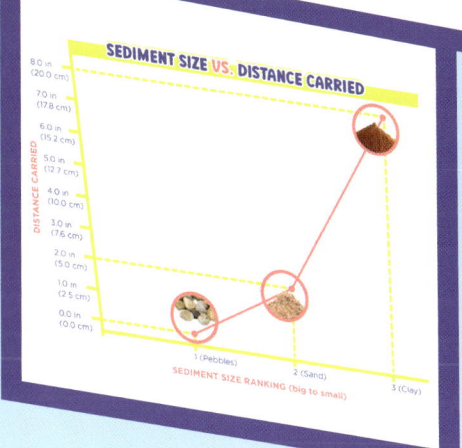

SEDIMENT SIZE VS. DISTANCE CARRIED

8.0 in
(20.0 cm)
7.0 in
(17.8 cm)
6.0 in
(15.2 cm)
5.0 in
(12.7 cm)
4.0 in
(10.0 cm)
3.0 in
(7.6 cm)
2.0 in
(5.0 cm)
1.0 in
(2.5 cm)
0.0 in
(0.0 cm)

DISTANCE CARRIED

1 (Pebbles)   2 (Sand)   3 (Clay)

SEDIMENT SIZE RANKING (big to small)

CONCLUSION

The results support my hypothesis that smaller sediment is carried farther downstream than larger sediment.

KEEP ASKING QUESTIONS

Your science project is over. You packed away your display. But don't stop asking questions! What might you do differently if you did the project again? What additional **research** could you do? Is there a related **topic** you would like to explore?

Beyond the Science Fair

Be a scientist beyond the science fair! You can use parts of the scientific method to find answers to everyday questions. Maybe you have a hypothesis for why rocks on a beach are smooth. Maybe you experiment to stop soil erosion in your garden. One day, you might use science to do big things. Maybe you'll help prevent river pollution! Turn your world into a science fair. What will you discover?

GLOSSARY

bias—showing a preference for one result over another.

design—to plan how something will appear or work.

focus—to concentrate on or pay particular attention to.

future—the time that hasn't happened yet.

information—the facts known about an event or subject.

mineral—a naturally occurring, solid substance that is not animal or vegetable, such as gold, ore, and some rock.

research—to find out more about something. Also, a study of something to learn new information.

summary—a short statement of the main points.

topic—the main idea or subject.

variable—a factor in a scientific experiment that may change.